Farm Kid's ABCs

Written By:
Jeanna Borgmann

Illustrated By:
Rebecca Yee

Copyright © 2025 by Jeanna Borgmann
All rights reserved. This book or any portion thereof
may not be reproduced or used in any manner
without the express written permission of the publisher
except for the use of brief quotations in a book review.

Written By Jeanna Borgmann
Illustrated By Rebecca Yee
Printed in the United States of America
ISBN: 978-1-7359107-4-1

A is for Acre.

An acre is how land is measured. One acre is a little less than the size of one football field.

Bb

B is for Birth.
A birth is when a baby animal is born. Sometimes animals need our help to safely give birth.

C is for Computers.
A computer is a machine that helps farmers use their equipment better. It can also keep checking the health of their animals and crops.

D is for Dairy.

A dairy farm is a farm that is made up of milking cows or goats. The milk can be made into other foods like butter, cheese, cream, yogurt, or even ice cream.

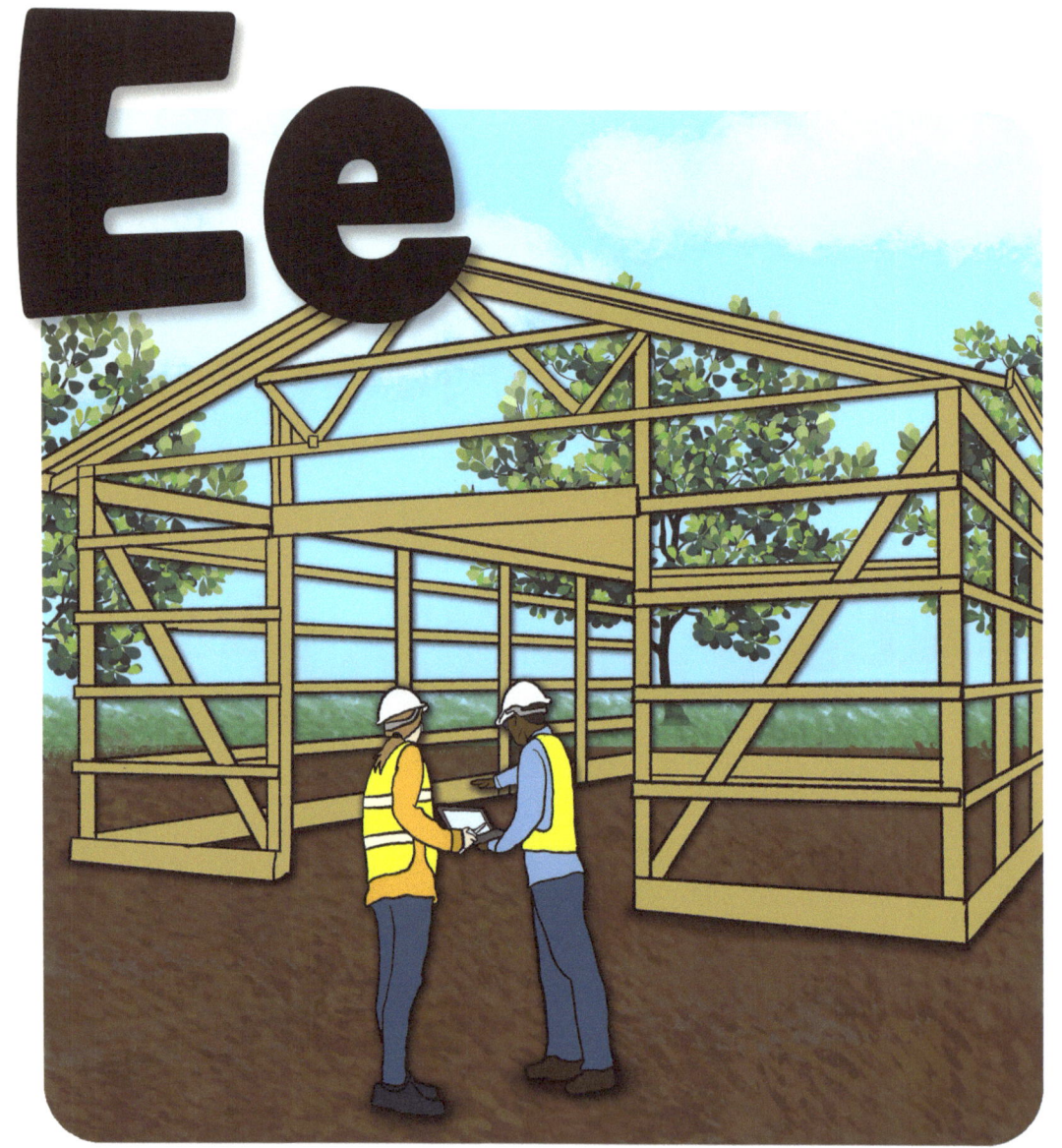

E is for Engineer.
Farmers must engineer, or design, build, and maintain, their farm to be sustainable and profitable.

F is for Family Farm.
A family farm is when the farm is run by a family like yours or mine. The majority of farms in the United States are family farms.

Gg

G is for Grazing.

Grazing is when cattle eat their food straight from the field or grassland. Similar to you or I taking food from a garden and eating it right away.

Hh

H is for Harvest.
Harvest is when farmers collect the crops off their fields. Some farms just have fields and no animals, these are called crop farms.

Ii

I is for Irrigation.

Irrigation is a water system for crops. It is similar to a sprinkler for a garden but bigger and with much more water.

J is for Jeans.

Jeans are the most durable pant that most farm families wear to work on the farm. Most farmers do not wear overalls anymore.

K is for Kid.
A kid is a baby goat. Goat farming is another common practice in the United States.

L is for Livestock.

Livestock refers to animals on a farm that work or produce for our consumption. For example, cows give milk for us to drink or make into other products.

M is for Meat.

Meat that we eat comes from animals. Farmers take good care of their animals so we'll have high-quality food to eat and it's their responsibility to treat animals well.

N is for Nourish.
Farmers must nourish their animals with foods that are rich in vitamins and minerals, so that they can grow and stay healthy.

O is for Operations.

Animals sometimes need operations, just like people, to keep them healthy. Animal doctors, called veterinarians, do those surgeries when animals need them.

P is for Piglet.

A piglet is a baby pig. Pig mothers usually give birth to at least 8-12 piglets at a time. But they could even have as many as 21 piglets in the same litter.

Q is for Quad Tractor.

A quad tractor is a farm vehicle with four tracks instead of wheels. Tracks spread out the pressure against the ground, helping keep farmers from getting stuck in the field.

R is for Robotics.

Robotics refers to using machines called robots to do jobs in place of a person. Robotics can be used for milking, pushing feed, scraping alleys, and driving tractors.

S is for Sand.
Sand is used for animal bedding. Animals like putting their toes in the sand too!

T is for Turkey Farms.
A turkey farm raises turkeys year round, not just during the holidays.

U is for Udder.
An udder is the part of an animal where the milk comes out.

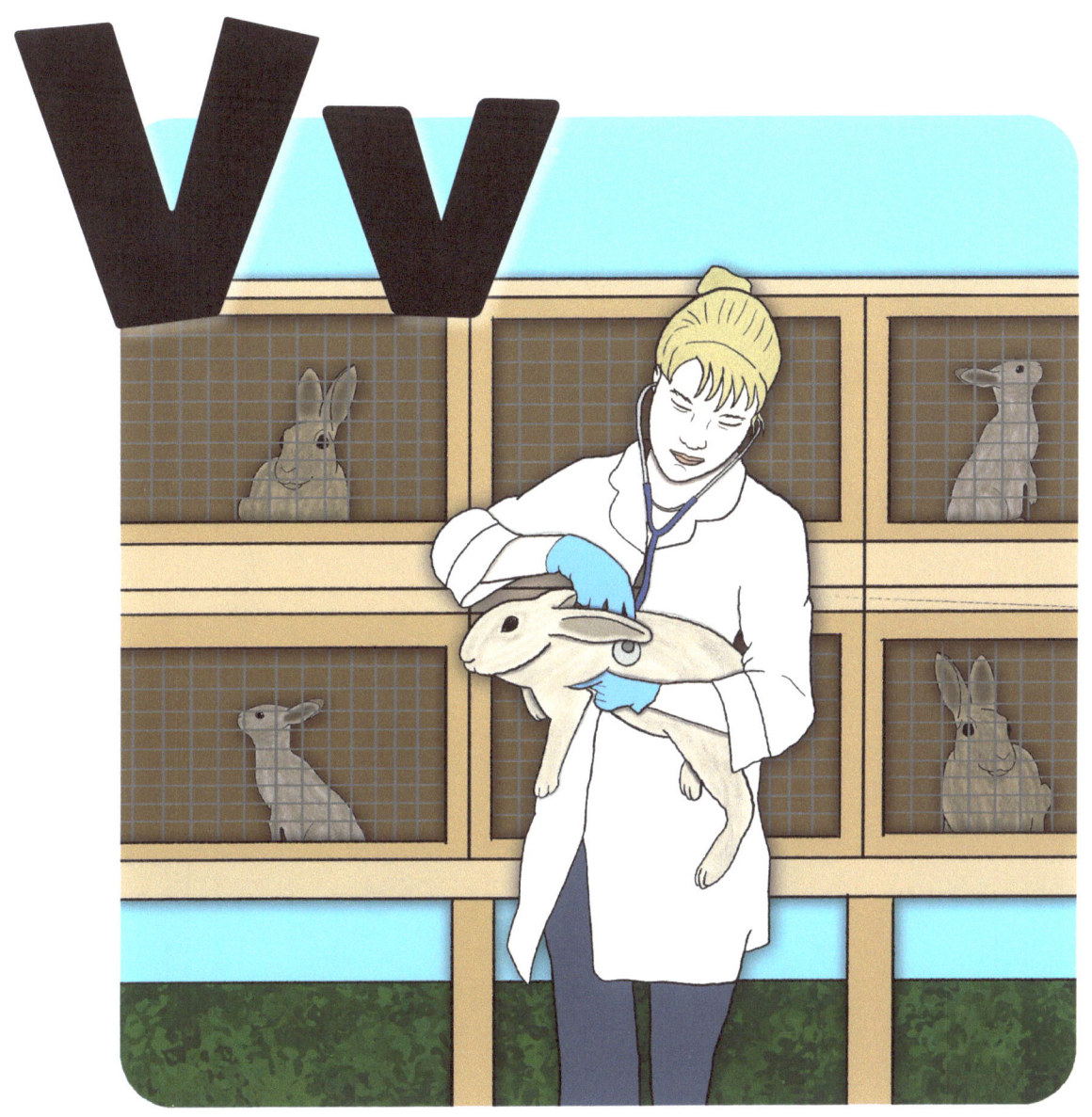

V is for Vet.

A vet is an animal doctor that helps keep the animals healthy. The word "vet" is short for "veterinarian."

W is for Weather.

The weather has a major effect on animal health and crop growth. Farmers need to be mindful of the weather and plan accordingly to best take care of their animals and crops.

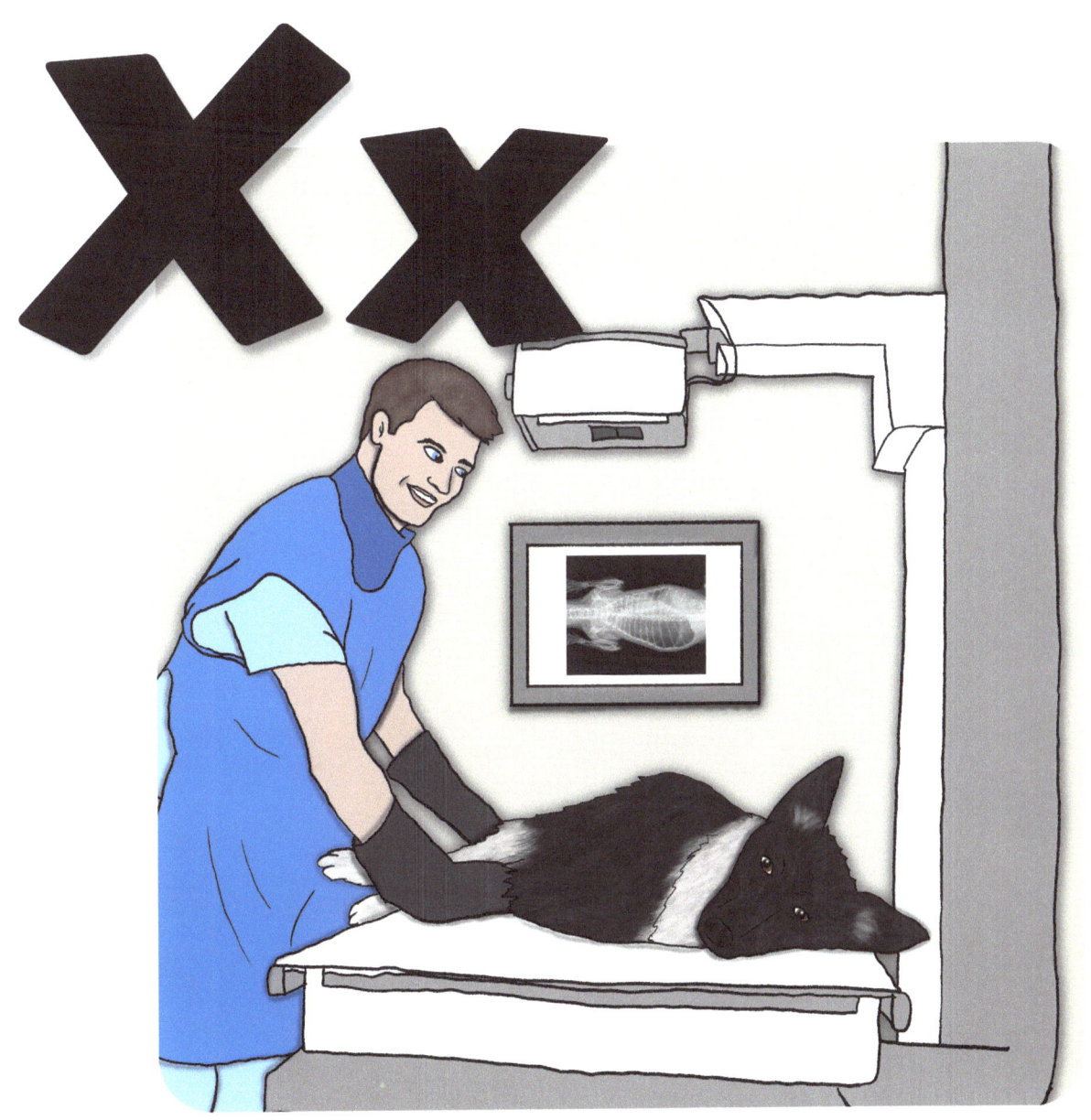

X is for X-Ray.

An X-ray is a special medical camera to take a picture of animals' bones, just like doctors of people use them.

Y is for Yearling.

A yearling is a year old animal. Similar to the toddler stage in people.

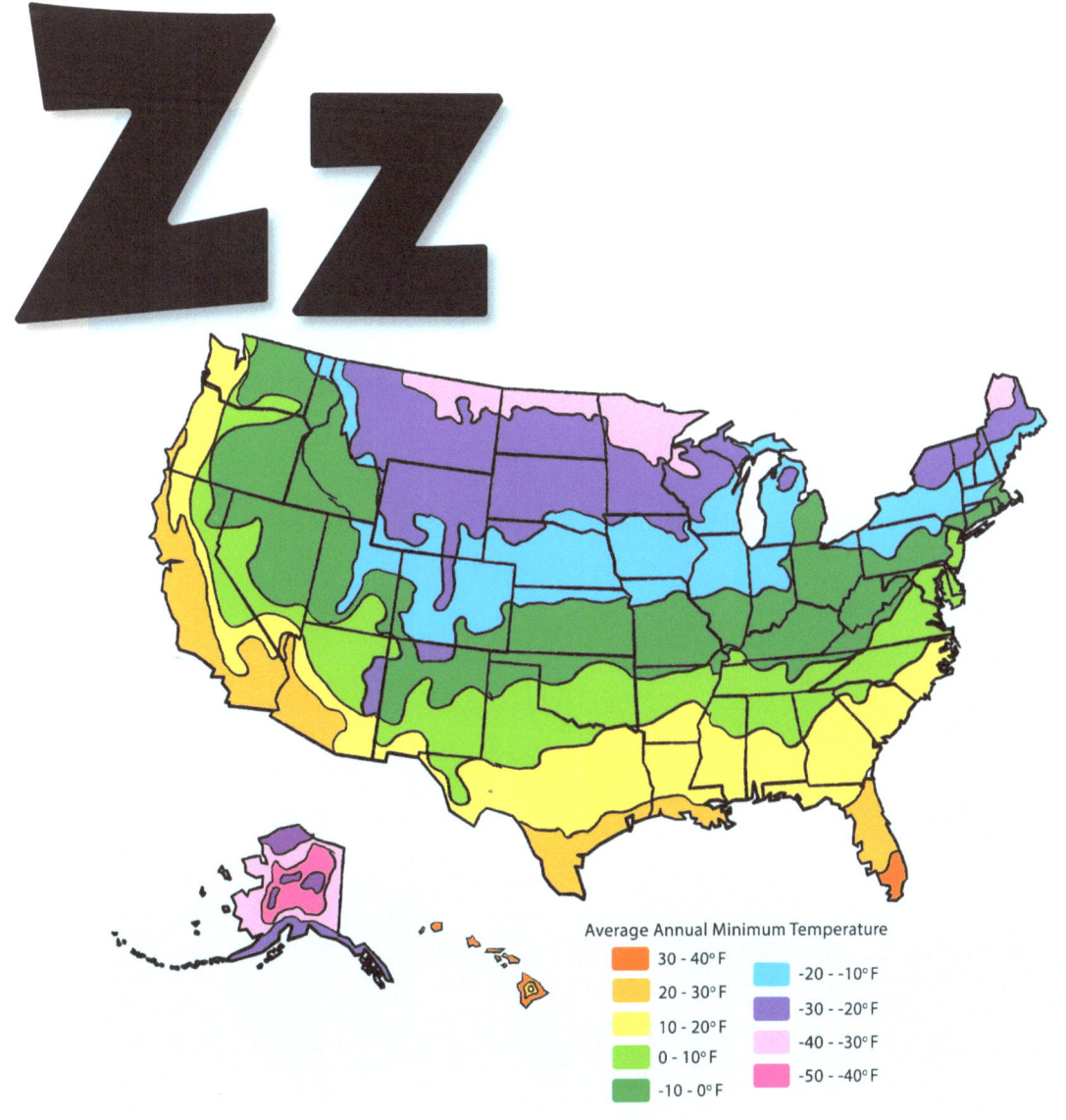

Z is for Zone.

A planting zone is an area of land known to be best for growing particular crops based on factors like average winter temperatures. It is also called a "hardiness zone."

About the Author

Jeanna Borgmann was raised on a small family farm in Central Minnesota, where a deep appreciation for hard work and rural life took root. Both she and her husband, Ryan, come from generations of farming families. Today, they live with their three daughters in the suburbs of the Twin Cities metro area, where Jeanna teaches middle school. Though she's no longer tending chores on the farm, those early experiences continue to shape her values and inspire her writing. Jeanna would like to extend a heartfelt thank you to the hardworking farmers—past, present, and future—who feed America every single day.

Check out other books from the author on Amazon:

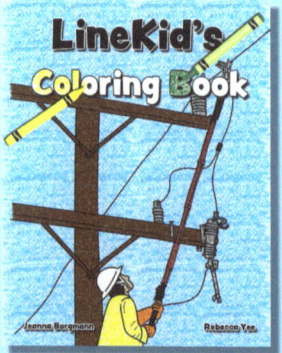

About the Illustrator

Rebecca Yee-Peters was born and raised in the small town of Massillon, Ohio, nestled near the edge of Amish country. From a young age, she developed a deep love for animals, spending her childhood caring for dogs, cats, and rabbits. She even showed rabbits at the county fair for many years. Today, Rebecca travels full-time in an RV with her husband, Eric, and their dog, Bailey, following his work as a Journeyman Lineman. Whenever they can, they return to help out on the Peters Family Farm, where dogs, horses and goats keep things lively.

You can explore more of Rebecca's work through *The Adventures of Pookie Entertainment* at AdventuresOfPookie.com.

If you enjoyed this book, please leave us a review online.

Reviews help indie authors like me reach a wider audience.

Thank you!

www.ingramcontent.com/pod-product-compliance
Lightning Source LLC
LaVergne TN
LVHW071029070426
835507LV00002B/79